FRUITS OF CREATION

A Look At Global Sustainability
As Seen Through The Eyes Of
GEORGE WASHINGTON CARVER

"And God said, Behold, I have given you every herb bearing seed, which is upon the face of all the earth, and every tree, in the which is the fruit of a tree yielding seed; to you it shall be for meat."
—Genesis 1:29

JOHN S. FERRELL

MACALESTER PARK
PUBLISHING COMPANY

mike Beard
800-407-9078

GREEN CROSS

Jeff says I should put this book on website. Need digital rights in protected environment

© 1995 by John S. Ferrell
Published jointly by
Macalester Park Publishing Company,
8434 Horizon Drive, Shakopee, MN 55379
and the Christian Society of the Green Cross,
10 East Lancaster Ave., Wynnewood, PA 19096
ISBN 1-886158-02-9

Permission requests should be directed to the author in care of the Christian Society of the Green Cross.

This book is published through grants from the Environmental Stewardship and Hunger Education program of the Evangelical Lutheran Church in America (ELCA) Chicago, Illinois, and the United Methodist Rural Fellowship, Columbia, Missouri.

Appreciation is expressed to Mr. Allen Johnson of the Green Cross for bringing this manuscript forward and seeing it through to publication.

Cover and text design: Smears Design, Minneapolis.
Printed and bound in the United States of America.

Photo credits:
George W. Carver National Monument
Diamond, Missouri

Tuskegee University
Tuskegee, Alabama

CONTENTS

PREFACE

One summer evening in 1990, I sat entranced as 97–year–old Wheeler McMillen vividly recalled a visit he had paid to George Washington Carver's laboratory more than sixty years before. McMillen, one of the most prominent agricultural journalists of his day, suddenly made real to me a man who had always seemed more like an image from *Ripley's Believe It or Not* than an actual historical figure.

But even before my encounter with McMillen, I had begun to realize that Carver was much more than a saintly scientist with a peculiar interest in peanuts. During the Great Depression, Carver, along with McMillen and other prominent Americans, joined together in a movement that advocated wide use of plant matter to produce fuel and industrial products. As a one–time renewable–energy activist, also trained in history, I was fascinated by this forgotten movement, which long before the oil crises of the 1970s, urged people to reduce their dependence on fossil fuels. For many months, I haunted the halls of the Library of Congress, trying to learn all I could about it.

It was this aspect of the Carver story that still had my attention in 1992, when I first spoke with Dr. Job Ebenezer, Director of the Environmental Stewardship and Hunger Division of the Evangelical Lutheran Church in America (ELCA). Job had recently returned from the Earth Summit in Rio de Janeiro. He saw an intriguing connection between Carver's work with poor farmers in the American South and Earth Summit discussions about environmentally sustainable development. Not only did he invite me to explore that connection, he also arranged to have the ELCA help with my research expenses. In a very real sense, Job is the father of this project, and I will always be grateful to him and to the ELCA for their support.

I am also grateful to William Shurtleff, Director of the Soyfoods Center in Lafayette, California. Bill was eager to learn more about a little known facet of Carver's multi–faceted career: his exploration of the soybean and its potential uses. I shared what little I found on this topic, and in return, Bill was generous in tapping the immense information resources of the Soyfoods Center to provide me with helpful background on Carver's life and his friendship with Henry Ford.

Finally, I am thankful for the constant encouragement and helpful suggestions of my wife, Jenny Holmes.

INTRODUCTION

In June 1992, delegates from more than 170 nations gathered in Rio de Janeiro, Brazil, for the United Nations Conference on Environment and Development (UNCED), commonly called the Earth Summit. At the same time, thousands of representatives from non–governmental organizations (NGOs) mounted a parallel, but less formal, Rio assembly to consider the same set of global challenges. It was a long and thoroughly daunting list that included deforestation, desertification, air and water pollution, ozone–layer destruction, global warming, hazardous wastes, loss of biodiversity, soil degradation, rapidly growing human populations, and persistent poverty for many of the world's people.

A principal outgrowth of the official Earth Summit was Agenda 21, a massive set of guidelines for worldwide action on Summit concerns through the rest of the decade and into the 21st century. Although the challenges addressed at the Summit and in Agenda 21 are complex, they can actually be summarized in a single question:

How can humankind assure that the future needs of all people are met within the context of ecological integrity?

Almost a century ago, working to help poor farm families in an environmentally damaged section of the American South, a remarkable educator began to confront that same question. A devout Christian as well as a profound student of nature, George Washington Carver saw that each part of the created world was connected to all the others and that everything God made had its message. To Carver, a waste was a misused resource and a weed was one of "God's plants for which we are not yet wise enough to have found uses."

Extending the Golden Rule to human stewardship of creation, Carver said farmers should be "kind" to the soil, and their kindness would be rewarded. To do otherwise, he believed, was to violate a basic principle. Nature, he warned, would drive away people who sinned against it.

In his early years at Tuskegee Institute, a school for blacks in rural Alabama, Carver demonstrated how even the poorest farm families could better their lives by restoring eroded soil and turning local plant and mineral resources to a host of unexpected purposes. He showed how under–appreciated plants, grown at home or gathered from the wild, could be used creatively to enhance both health and self–reliance, and he taught how these benefits could be extended throughout the year with simple food preservation techniques.

Recognizing that the poor were nourished by more than food, Carver showed how colors from local clays could beautify their homes and how local trees, vines, and shrubs could enhance the appearance of "every park and dooryard."

In the last part of his long career, Carver extended his self–reliance philosophy well beyond the family farm by demonstrating how local renewable resources could be used to build new regional industries. He made synthetic marble from wood shavings and found processes to turn plant matter into myriad new consumer products. From peanuts he produced face cream, insecticide, and plastic. From sweet potatoes he produced paints, ink, and an experimental rubber compound. He also pointed hopefully to the potential for producing alcohol fuel from farm crops.

Carver's contemporaries, grappling with the Depression of the 1930s and the wartime shortages of the early 1940s, understood, to a limited degree, that his work with local renewable resources was important. But his fame was based in part on mythology. He was portrayed in the press as a great scientist whose discoveries had done much to transform the southern economy. In reality, he made no great scientific breakthroughs, and few of his discoveries were commercialized in his own day or later.

After his death, an affluent America mesmerized by petrochemical consumerism could not relate to a figure who warned about wastefulness and advocated better use of renewable resources. Carver was largely relegated to juvenile biographies as a kindly old man who did "something with peanuts."

But Carver's real importance cannot be measured by the standards of his own day or ours. In his life and thought are seeds of currently emerging trends that many view as harbingers of a sounder future: organic agriculture, renewable energy, "green" consumer products, better waste utilization, environmental restoration, and regional self–reliance.

Carver's talents and involvements were remarkably diverse, and his story has been told many times. This book, which relies to a large extent on his own words and those of his contemporaries, focuses on aspects of Carver's life that made him a "blazer of trails" to a sustainable future—one in which each generation of humankind will achieve its well–being in ways that protect the earth and its ecosystems, leaving them intact and fruitful for all generations to come.

To my wife,
Jenny Holmes

"MY VERY SOUL THIRSTED FOR AN EDUCATION"

Neighbors knew there was something unusual about the frail little black boy. They called him the "Plant Doctor." As an adult, he recalled that "sick plants were brought by the score, and left for treatment, and I often went to houses, and prescribed for them, much as a physician prescribes for his patients."[1]

It was a few years after the close of the Civil War, and George was growing up in the home of Moses and Susan Carver, who had a prosperous farm in southwestern Missouri. George did not know the year of his birth (although it was probably 1864 or 1865)[2] and he never knew his real parents. His mother had been a slave of Moses and Susan, and he was told that his father was a neighbor's slave, who had died in an accident shortly after George was born.[3]

As a slave state that chose to stay in the Union, Missouri was a hotbed of conflicting political sentiment during the war. The area in which George was born was plagued by bands of

Confederate bushwhackers, Union raiders, and common outlaws. It was apparently one or another of these bands that swooped down on the Carver farm and kidnapped the infant George and his mother, carrying them off to Arkansas. John Bentley, a Carver neighbor, mounted a search, but was only able to recover little George. The story became a favorite of journalists and biographers in later years, and depending on who was telling the story, Moses Carver either rewarded Bentley with a race horse or Bentley used the horse to ransom George in Arkansas.[4]

After the war, George and his older brother Jim became, in effect, the Carvers' adopted sons. Although Moses and Susan were not churchgoers, young George found his way into the Sunday school of a local white congregation. Lacking a pastor of its own, the congregation depended on visiting clergy from a variety of denominations.[5] Christian faith was to play a central part in the George's later life, but like the little church where his spiritual training began, he would limit himself to no single denomination.

George, at an early age, discovered his love of knowledge and his affinity for nature. "My very soul thirsted for an education," he recalled as an adult. "I literally lived in the woods. I wanted to know every strange stone, flower, insect, bird, or beast. No one could tell me."[6]

He was excluded from the local white school, and the only book available to him at home was

14

an old Webster's elementary spelling book, which he learned almost by heart. Finally, around 1877, Moses and Susan let him move to Neosho, a town some eight miles from home, where he was able to attend classes at a school for blacks while living with a local family. From Neosho, he moved from community to community in Kansas, continuing his education and supporting himself by performing a variety of jobs, including household chores for local families. Briefly, he was employed in a clerical position at the Union Depot in Kansas City.[7]

Around 1885, George Carver, by now a young man, was accepted by mail at a small Presbyterian college in Highland, Kansas. When he arrived to register, it was a different story: he was informed that the school did not accept students of his race.

His hopes crushed, Carver worked for a time in Highland, then resumed his life of drifting. Eventually, he found himself in Winterset, Iowa, where he found a job as a cook in a hotel. Decades later, he described how a chance encounter in the Iowa town set in motion a series of events that dramatically changed his life:

> **One evening I went to a white church, and set [sic] in the rear of the house. The next day a handsome man called for me at the hotel, and said his wife wanted to see me. When I reached the splendid residence I was astonished to recognize her as the prima dona [sic] in the choir.**

Fruits of Creation

> **I was most astonished when she told me that my fine voice had attracted her. I had to sing quite a number of pieces for her, and agree to come to her house at least once a week....**

The woman discovered Carver's fondness for painting, and impressed by his artistic abilities, she and her husband urged him to apply to Simpson College, a small Methodist school twenty miles away. He did so and was admitted. Beginning his studies at Simpson in the fall of 1890, Carver supported himself by operating a laundry.[8]

Etta Budd, Carver's art teacher at Simpson, was impressed by his talent, but knowing of his love for plants, she suggested that he transfer to the Iowa State College of Agriculture and Mechanic Arts at Ames, where her father was a professor of horticulture. Carver decided to follow her advice, and left for Iowa State in 1891.[9]

He thrived at Ames, where he was well accepted by the student body and was active in student religious life. For a time, he was part of a men's prayer group that met each Wednesday evening in the office of James Wilson, dean of agriculture.[10] Carver was close to Wilson, and when Wilson later became U.S. Secretary of Agriculture, he remembered Carver with favors and encouragement.

After he received his bachelor's degree in agriculture, Carver was appointed to the faculty as an assistant in botany and given charge of the school's greenhouse. He also began to pursue his master's degree at the school, and gained recognition for his growing skills in mycology, the study of fungi.[11]

In 1896, as he approached the end of his graduate studies, Carver seemed poised for a distinguished, but conventional, scientific career. But then he received a letter that would radically alter the course of his life.

Study Questions
Chapter 1

Even as a boy, Carver enjoyed spending time in the woods. He was filled with curiosity about the stones, flowers, insects, birds, and animals he found there.

What kinds of things can we learn from careful observation of creation?

How did Carver use his knowledge of plants to help his neighbors?

What have you learned while engaging in such outdoor activities as hiking, camping, or gardening?

Although clearly gifted and blessed with loving adoptive parents, Carver had to overcome serious barriers to excel in his chosen fields.

What were the barriers Carver had to overcome?

What personal qualities allow a person to rise above the kinds of barriers Carver encountered? How does a person develop such qualities?

After Carver was turned down for college admission because of his race, a white couple he met through a chance encounter at a church service befriended him, saw his artistic talent, and urged him to apply to another school.

What responsibility do we have to encourage others— including the disadvantaged—to develop their talents?

Do you know of other cases where someone with poor prospects in life has benefited by the kind of encouragement Carver received? Have you personally benefited?

Can you think of ways that churches could do more to help young people see and develop their potential?

"TO BETTER THE CONDITIONS OF OUR PEOPLE"

T he letter was from Booker T. Washington, then one of the most influential black leaders in America. A former slave, Washington had begun in 1881, with practically no resources, to build the Tuskegee Normal and Agricultural Institute, an impressive school for blacks in rural Alabama. Washington believed strongly in self–help: blacks in the South helping other blacks to gain an education, learn trades, become better farmers, and generally improve the quality of their lives. He was criticized, with some justice, for being an accommodationist, willing to acquiesce in some southern racist political strictures to gain white support for his self–help program. Nevertheless, in carrying out that program, he influenced the course of many lives for the better,[1] and in George Washington Carver, he found a self–help genius to aid him in his task.

Carver's special gifts were not fully evident to Washington. The Tuskegee principal knew only that he wanted to establish a full–scale department of agriculture at Tuskegee and that Carver appeared well qualified to head such a department.

At first, Carver was ambivalent. He could stay at Iowa State, where he was held in high esteem, or he could accept a tempting offer then being made by another school. Finally, he agreed to Washington's offer, telling the Tuskegee principal of his intention to cooperate with him "in doing all I can through Christ who strengtheneth me to better the conditions of our people."[2]

Carver may have underestimated just how much strength he would need. Decades later, he described what he had first encountered in the rural districts around Tuskegee: "devastated forests," eroded soil, poorly prepared land, no crop diversification, very little livestock, lack of home gardens, and poor family diets.[3]

The deforestation and soil erosion had much to do with "King Cotton," which had for so long dominated the regional economy. Once cleared, land had been planted to cotton year after year, and when Carver arrived in the South, this destructive obsession was being reinforced by a dilemma that typically faced poor farmers, many of them blacks only one generation removed from slavery. Max Bennett Thrasher, an associate of Booker T. Washington, described this dilemma:

In many cases the farmer owns no land, and

as a result must rent a "patch" on such terms as he can make with the landlord, to whom he contracts to deliver a certain portion of the crop for rent. He has little or no money with which to purchase supplies in advance, and so, before his crop is even planted, he has to mortgage the balance of it to some merchant in town for food for himself and family to live on through the spring and summer. As cotton is the readiest cash crop in the country, neither landlord nor merchant wishes to make an advance on any other crop. As a result, the farmer too often is forced to plant only cotton—buying even his corn meal and bacon of the storekeeper, and of necessity obliged to pay almost any price which the dealer may demand.[4]

Clearly Washington, in his early years at the school, faced a challenge in reaching out to southern blacks who faced such conditions in their daily life. The appalling physical aspects of that challenge were well described by Thomas Monroe Campbell, who began as a student at Tuskegee only three years after Carver joined the faculty: "Not only near the school, but throughout the Black Belt in Alabama and other Southern States could be seen hundreds of squalid, ramshackled cabins, tenanted by forlorn, emaciated, poverty stricken Negroes who year after year struggled in cotton fields and disease–laden swamps, trying to eke out a miserable existence."

Pig pens, Campbell noted, were often at the very doors of these dwellings, and wells "were in many instances down the hill from these pens, or close by them." Many of the homes lacked toilets, and "where there were, they were often just a few boards nailed crudely together with a piece of sack serving as a door."[5]

In such surroundings, the Tuskegee campus stood in sharp and exhilarating contrast. Campbell remembered the strangeness of the scene he encountered upon his arrival: "...sawmilling [and] brick–making; the construction of houses, carriages, wagons and buggies and the making of tin utensils, harness, mattresses, brooms, clothes, shoes—all done by Negroes...." Campbell, schooled in the assumption that work other than farming could only be done by whites, was overwhelmed. It was, he recalled, "like entering a new heaven, and I could scarcely believe that such things were possible."[6]

Booker T. Washington's hope was that his students, after leaving Tuskegee, would return to their people in a sort of missionary role, spreading enlightenment and serving as examples to blacks who had not had the same educational advantages.[7] However, he also realized the need for more direct outreach efforts by the school. Out of this realization grew an imaginative extension program to which Carver would lend his own unique approach to self–help through better understanding of Nature's gifts.

Study Questions
Chapter 2

Even though his career prospects in Iowa were good, and even though he had a job offer from another school, Carver chose Tuskegee.

What motivated him to make this decision? What does Carver's decision say about him as a person?

What would you have done if you had been faced with Carver's choices?

When Carver arrived in Alabama to begin his work at Tuskegee, he found appalling conditions near the school: eroded soil, ruined forests, and people living in squalor as they struggled to make a living from the land.

To what extent was environmental damage a factor in the problems facing the people Carver saw?

Why were people growing so much cotton and so little food?

What similarities do you see between the conditions Carver saw near Tuskegee and conditions in developing countries today? What differences?

Thomas Campbell recalled that when he arrived at Tuskegee, it was "like entering a new heaven, and I could scarcely believe that such things were possible."

What did Campbell see that so impressed him?

Why was it so difficult for him to believe that such things were possible?

What did Booker T. Washington hope that Tuskegee students would accomplish when they left the school?

"BEING KIND TO THE SOIL"

I n 1897, thanks to Booker T. Washington's lobbying skills, the governor of Alabama approved an act establishing the Tuskegee Agricultural Experiment Station for "educating and training colored students in scientific agriculture." Under Carver's direction, it became the first all–black experiment station in the United States, but the funding provided by the state for its operation was paltry. As Carver biographer Linda O. McMurry has noted, "Other experiment station staffs included separate chemists, botanists, entomologists, and mycologists. At the Tuskegee station all these positions were filled by one man—Carver."[1]

Fortunately, that one man was a generalist by nature as well as necessity, and in his operation of the experiment station, he always remembered the people most in need of his help. As he noted in one of his station bulletins,

For eight years the Tuskegee station has made the subject of soil improvement a special study,

emphasizing the subject of crop rotation, deep plowing, terracing, fertilizing, etc., keeping in mind the poor tenant farmer with a one–horse equipment; so therefore, every operation performed has been within his reach, the station having only one horse.[2]

In keeping with Booker T. Washington's self–reliance philosophy, Carver was attempting to show poor farmers how to break the cycle of debt by adopting effective, inexpensive ways to increase their production. He also encouraged families that were able to do so to plant gardens and preserve some of what they grew for the winter months. In that way, they could supply more of their own needs and save scarce cash formerly wasted at the local store. Additional cash could be raised by selling some of the produce, and ultimately, a farm family could put into practice what Washington called "the gospel of Tuskegee": that every black family should own its own farm or home.[3]

In his early years at Tuskegee, Carver would sometimes load some tools and exhibits into a buggy and set out on a weekend trips to rural areas near the campus where he gave demonstrations designed to help farm families improve their lives. He favored Sunday demonstrations because he could find ready–made audiences gathered for worship at their local churches.

Booker T. Washington saw the value of this form of outreach, and in 1904 he proposed the idea of a farmer's school on wheels that could travel into the countryside. Carver responded that "your idea is a most excellent one. Germany, Canada and other countries, I understand, do this with success." He presented Washington with a rough sketch of how the school wagon might be designed, recommended equipment for it to carry, and suggested a variety of demonstrations, for both farmers and their wives, that the school could carry out in the field.[4]

Washington appointed a committee headed by Carver to draw up plans for the wagon, and a year and a half later, the Tuskegee traveling school began its trips into the countryside. Soon after, Washington made an arrangement with the U.S. Department of Agriculture, which had only recently initiated its own farm demonstration program, to serve as a sponsor of the project. Thomas Campbell, by then a Tuskegee graduate, was selected as operator, and a token one dollar a year of his salary was paid by the U.S. government. Campbell thus became the first black agricultural demonstration agent in federal employ. The traveling school, later mounted on a truck, continued to operate until World War II.[5]

Carver experimented with a number of crops in his early years at Tuskegee, placing special emphasis on those that had soil–building qualities, versatile uses, or special potential for breaking cotton's economic grip. Although agricultural specialists were already advocating commercial fertilizers, Carver pointed to the value of nitro-gen–fixing plants, such as cowpeas and velvet beans, and to the benefits of applying locally available natural fertilizers. In a 1916 article he wrote that three acres of the Experiment Station had not had commercial fertilizer applied to them for twelve years. Instead, a compost, comprised of two–thirds leaves from the woods and muck from the swamp and one–third barnyard manure, had been used. ("Muck," he explained, was "simply the rich earth from the swamp.")

Carver described how to make such a compost and use it together with wood ashes and waste lime. He assured farmers that "It will take only one or two trials to convince you that many thousands of dollars are being spent every year here in the South for fertilizers that profit the user very little, while Nature's choicest fertilizer is going to waste."[6]

This interest in improving both soil and profits by utilizing overlooked resources was still very much in evidence late in Carver's life. In 1938, he described a successful experiment begun the previous year at Tuskegee to turn paper, leaves, rags, grass, and other trash into compost. When the pile was about two feet thick, it was covered over with "rich earth from the woods and swamps." The compost pile, he reported, "has saved the school $250." He concluded that

If farmers of the South would unite (observing every sanitary rule with reference to dis-

eases of any kind, as well as fungus troubles
and insect enemies) building a large or small
[compost] pen according to their needs, and saving all of this valuable fertilizing material,
returning same to the soil, it would not be long
before the South would have but few if any
non–productive areas.[7]

Carver saw that good treatment of the soil and
loving human relationships had something very
much in common. In a 1914 magazine article
entitled "Being Kind to the Soil," he observed
that

Unkindness to anything means an injustice
done to that thing. If I am unkind to you I do
you an injustice, or wrong you in some way. On
the other hand, if I try to assist you in every way
that I can to make a better citizen and in every
way to do my very best for you, I am kind to you.

The above principles apply with equal force
to the soil. The farmer whose soil produces less
every year, is unkind to it in some way; that is,
he is not doing by it what he should; he is rob-
bing it of some substance it must have, and he
becomes, therefore, a soil robber rather than a
progressive farmer.[8]

To underscore his contention that farmers
should turn away from over–dependence on cotton, Carver turned in a 1921 lecture to a soil man-
agement model from the Bible:

We take this very Book, here—go way back
here, almost to the beginning of time and we

find, way back in the time of the Pharaohs, the farmers were obliged to rest their lands and every fifty years was jubilee year. This was picnic time for the soil. Nothing must be taken off of it. Everything it produced was to go back to the soil. Now then, you know as well as I know that ever since that time we have heard such terms as diversify, diversify, diversify—rest your soil. We paid absolutely no attention to it.[9]

On a trip to Georgia many years later, Carver displayed his exasperation with the ecological blindness he still saw about him. He was quoted as follows in an Atlanta newspaper:

Conservation is one of our big problems in this section. You can't tear up everything just to get the dollar out of it without suffering as a result.

It is a travesty to burn our woods and thereby burn up the fertilizer nature has provided for us. We must enrich our soil every year instead of merely depleting it. It is fundamental that nature will drive away those who commit sins against it.[10]

Study Questions
Chapter 3

Carver taught impoverished families how to supply more of their own needs and become more economically self-reliant.

In view of the laws that restricted opportunities for blacks in the South of Carver's day, was his self-help strategy sufficient? Why didn't he take a more political approach, urging people to challenge those laws?

If Carver were alive today, how do you think he might adapt his self-help strategy to better the economic situation of people living in inner-city neighborhoods?

Carver wrote about the benefits of "being kind to the soil."

What did kindness to the soil mean to Carver? What does it mean in today's world?

What connection did Carver see between kindness to the soil and kindness to other people?

Do you think Carver would consider current farming practices—such as heavy use of pesticides—to be consistent with kindness?

In a newspaper story, Carver was quoted as saying that "It is fundamental that nature will drive away those who commit sins against it."

What do you think Carver meant by this?

Can you point to examples in our own day of nature turning against people who have "sinned against it"?

Do you think "sin" is the appropriate word to use for our abuses of God's creation?

"A GREAT TEACHER"

More than four years before Carver's arrival at Tuskegee, Booker T. Washington sent out about 75 invitations to farmers, mechanics, school teachers, and ministers to spend a day at the school discussing their condition and needs. To his surprise, several hundred men and women showed up. They discussed shared challenges, such as dealing with debt and providing adequate educational opportunities to their children. Then they turned to possible solutions and adopted a resolution that included this prescription for bettering their own lives and the lives of people like them:

> We urge all to buy land and to cultivate it thoroughly; to raise more food supplies; to build houses with more than one room; to tax themselves to build better school houses, and to extend the school term to at least six months; to give more attention to the character of their leaders, especially ministers and teachers; to keep out of debt; to avoid lawsuits; to treat our women better; and that conferences similar in

aim to this one be held in every community where practicable.[1]

Washington was so excited by the success of this first conference that he made it an annual event. Inspired by the Tuskegee example, other southern black educational institutions organized similar programs, and hundreds of local conferences sprang up as well.[2]

Soon after Carver became head of the new Tuskegee agricultural department, the school initiated additional outreach programs. Beginning in 1897, it offered Farmers' Institutes on campus. One day each month, farmers received free training in the basics of good agriculture. The institutes, in turn, spawned an annual fair at Tuskegee where farmers and their families could display their crops, livestock, needlework, quilts, and canned goods. The fair eventually expanded into a county–wide event, which Carver described as a source of "strength, information, inspiration, and encouragement."[3]

At Carver's urging, a "Short Course in Agriculture" was initiated at Tuskegee in 1904 to provide farmers with practical instruction during the winter season, when they had time on their hands. Beginning as a six–week program, the Short Course evolved into an annual two–week affair. Since the program included special classes for women and children, whole families would often attend.[4]

Carver served as one of the instructors of the Short Course. Despite his oddly high–pitched voice, he had remarkable skills as a speaker, and he put them to frequent use, not only in extension courses and his regular Tuskegee classes, but also at farmers' conferences held elsewhere in the South.

Carver even led a popular Sunday–evening Bible class at Tuskegee for thirty years. One participant commented on his "remarkable" teaching method, noting that "He brings the Bible characters home to the students by impersonating them, and shows conclusively that there is no possible conflict between science and the Bible."

Booker T. Washington recognized Carver's special gifts on the podium, calling him "a great teacher, a great lecturer, a great inspirer of young men and old men...." A biographical sketch from 1918 noted that Carver's "name attached to a placard or bulletin announcing a proposed farmers conference will draw a larger number of interested individuals—both white and black—than the name of any other speaker who may be secured."[5]

Carver knew that if he was to have an impact on listeners who were illiterate or barely educated, he would need to communicate his profound respect for the natural world in basic language. The wholeness of that world, so evident to him, would have to be made evident to his listeners as well. As he wrote in 1902,

> Let farmers' institutes be organized, and all the methods of nature study be brought down to the everyday life and language of the masses. Let us become familiar with the commonest things about us, of which two–thirds of the people are surprisingly ignorant. The highest attainments in agriculture can be reached only when we clearly understand the mutual relationship between the animal, mineral, and vegetable kingdoms, and how utterly impossible it is for one to exist in a highly organized state without the other. If every farmer could recognize that his plants were real, living things, and that sunshine, air, food, and drink were just as necessary for their lives as for that of the animal, the problem would become at once intellectual, enjoyable, and practical.

How well he succeeded in imparting this message may be gauged from this report of his visit to a 1915 Farmers' Conference at Albany Institute in Georgia:

> Prof. Carver is a genius. He not only knows his subjects but puts them in such simple form that a child can grasp them. His knowledge of the soil and plant life is simply wonderful. No one can spend any time with Prof. Carver in a grove or woods without getting some conception of nature and nature's God. He sees the good and beautiful in everything that God has made.[6]

Not leaving his message solely to his formidable verbal skills, Carver also made frequent use of visual displays. Describing a Carver exhibit mounted at a fair in 1915, one writer told how the professor "showed by actual results" how farm families could use items they grew themselves to live comfortably through the winter:

> **His showcase was filled with home–grown products cooked and prepared temptingly in a dozen different styles. In the jars upon the shelves in his booth there were the same home–grown products dried, canned, pickled and preserved, and to cap it all he served a cup of delicious coffee made with dried kidney beans, and treated in some simple fashion. Commenting on Professor Carver's booth, the Montgomery *Advertiser* says, "To see which makes one wonder why he has been so blind to their use for so many years.[7]**

The Tuskegee Experiment Station published a series of bulletins on subjects as diverse as "How to Build Up Worn Out Soils," "Saving the Wild Plum Crop," "Increasing the Yield of Corn," and "Successful Yields of Small Grain." In addition to numerous articles in local and national periodicals, Carver wrote most of the bulletins himself.

In the first of the bulletin series, published in 1898, he was frank about his intention to use information from bulletins of other stations when it seemed pertinent. In fact, information in a number of the subsequent bulletins was not new for people with the means to find it. Decades after his death, critics, in reassessing Carver's reputation as a "great scientist," pointed to this lack of originality. They also noted that some of the uses for the peanuts he would later publicize had been described earlier by others.[8] But Carver's real greatness lay not in original research, but in his ability to see and understand the potential gifts inherent in creation and describe those gifts in a way that even the uneducated could understand.

Sensitive to the needs of many of his readers, he pledged in the first bulletin that "few technical terms will be used, and where such are introduced, an explanation will always accompany them." This goal of writing plainly was still evident in one of his last bulletins, *Nature's Garden for Victory and Peace*, published in 1942. Carver explained to a correspondent that

> **It is written just as simply as I could do it, as I was trying to have it reach the great mass of people....I have left out all technical terminology with the hope that it would serve those people better as the bulk of the people it is intended to help most know absolutely nothing about vitamins. In fact they never have and they never will. They can understand when they are hungry that certain wild vegetables are edible.[9]**

Good teacher that he was, Carver recognized that education was a two–way process. He said he always liked "to hear the common folk talk, because you can get a good deal of information...if you will allow them to talk and you listen."

Carver's Tuskegee colleague G. Lake Imes told

the story of the professor's encounter with an elderly lady who was the mother of another faculty member. Seeing some dried herbs hanging in Carver's laboratory, she mentioned certain household remedies she knew about. Carver was immediately alert, and the two conversed for hours. Afterward, Carver "averred that this modest, obscure old lady knew more about the medicinal value of a wider range of herbs and plants than anyone he had ever met and that he had learned a lot from her."[10]

Carver believed that people learned best by beginning with something they already knew, then proceeding to the nearest related unknown. Addressing "those who have not yet learned the secret of true happiness, the joy of coming into the closest relationship with the Maker and Preserver of all things," he gave this advice: "[B]egin now to study the little things in your own door yard, going from the known to the nearest related unknown, for indeed each new truth brings one nearer to God."

A journalist wrote Carver that "In a large measure you have reorganized my thinking and outlook at life. I especially got a new slant on effective living when I had absorbed your idea of education, when you say that 'education is understanding relations.'"[11]

Carver was very aware that the knowledge of every member of the family was important to the well–being of individual farms and rural communities. "Attend with your family every farmers' meeting or conference," he urged farmers in an 1899 article. He further advised them to send their sons to "the best agricultural school within your reach" and to let their daughters "learn the technique of poultry raising, dairying, fruit growing, and landscape gardening."[12]

In 1910 Carver published an Experiment Station bulletin entitled *Nature Study and Gardening for Rural Schools*, which included practical exercises to help teachers acquaint children with the wonders of the natural world. In the introduction to the bulletin, he expressed his "hands–on" philosophy:

> **The thoughtful educator realizes that a very large part of the child's education must be gotten outside of the four walls designated as class room. He also understands that the most effective and lasting education is the one that makes the pupil handle, discuss and familiarize himself with the real things about him....[13]**

During his youth and young manhood, Carver had often supported himself doing cooking and other domestic chores usually reserved, in the social arrangements of that day, for women. With this background, he could well appreciate that reaching women was critically important in his efforts at Tuskegee to help rural families become more self–reliant. In the backward environment of rural Alabama, he found that "the average housewife knew but little about the nutritive value of foodstuffs, economy in selection, and palatable and attractive preparation."[14]

In a 1909 article, Carver urged farmers to sit down with their wives and plan a diversified agricultural program aimed at greater prosperity and self–reliance. In Experiment Station bulletins, he included recipes to provide farm women with creative ideas for preparing meals based largely on ingredients grown or gathered at home rather than purchased with scarce cash.

To make food self–reliance a year–round proposition, Carver also stressed the importance of canning and drying. In 1916, a writer profiling the Tuskegee educator noted how

> his experience in cooking, coupled with his knowledge of chemistry, enabled him to discover methods of preserving hundreds of dollars' worth of fruits and vegetables that had before been thrown to the hogs or allowed to rot on the highway. He taught the farmers' wives how to preserve the cow pea when it was in the green pod and how to vary the cooking of it from day to day....He showed how to make wild–plum syrup, plum vinegar, plum duff, plum preserves, and how to convert green plums into olives. The results of this teaching enable many a farmer's wife in Alabama today to point to her preserve closet with pride.[15]

Study Questions
Chapter 4

According to Booker T. Washington, Carver was "a great teacher, a great lecturer, a great inspirer of young men and old men...."

What methods did Carver employ to make his educational efforts effective?

How did he approach the challenge of bringing information to people whose education was limited?

How did he benefit from listening to "common folk"?

A writer observed that Carver saw "the good and beautiful in everything that God has made."

Why was he able to do this?

Was he simply overlooking "the bad and the ugly" in the world around him?

If Carver could visit one of our inner-city neighborhoods, would he see examples there of the "good and beautiful" in creation?

In addressing people who had "not yet learned the secret of true happiness, the joy of coming into the closest relationship with the Maker and Preserver of all things," Carver advised them "to study the little things in your own door yard, going from the known to the nearest related unknown...."

What benefit did Carver say "each new truth" uncovered in such study could bring?

If you were to "study the little things in your own door yard," what would you start with? What would be "the nearest related unknown"?

Do you think Christians could make use of this method of learning to better understand nature and their own roles as stewards of creation?

"WALKING ON ROSES"

Late in her life, Etta Budd, Carver's art teacher at Simpson College, recalled that her unusual student had been fond of color, design, and form. He could draw anything.

Although he decided rather reluctantly to pursue another career, Carver never stopped being an artist. More than a decade after arriving at Tuskegee, he could still say, "I am an artist by taste, training, and profession; therefore, nothing pleases me more than to take my pencils, paint, brushes, sketch book, pastel board, etc., and spend a day in the woods."[1]

His mission to black farmers extended beyond the practical requirements of daily life, because he knew, as an artist, that improved diets and increased economic self–reliance were not enough for poor farm families. They also needed the nourishment of natural beauty.

In an Experiment Station bulletin, Carver noted that among the plants in his part of

Alabama one could find "flowers of rare beauty and fragrance, foliage unsurpassed in richness, and fruits, berries and other forms of seed capsules possessing a richness of color and gracefulness of form, which well nigh approaches the ideal in beauty and grace." Such plants, free for the taking, could be used to beautify one's yard, and, as he noted in another bulletin, flowers were "soothing and restful to the tired body and brain."[2]

"I never saw anyone love flowers as Dr. Carver does," wrote his friend Glenn Clark. Another friend, Tuskegee colleague G. Lake Imes, recalled that it was Carver's practice to "wear in the lapel of his coat a native flower of some kind, picked by the wayside, the more obscure the better, which he wore with the same pride and satisfaction that milady derives from a corsage of orchids." The contrast between the flower's beauty and the characteristic shabbiness of the coat made his aesthetic priorities obvious.[3]

In another of his bulletins, Carver described how native clays could be used to produce white and color washes that would improve the appearance of humble homes, both inside and out. The bulletin was primarily designed, Carver said, "to aid the farmer in tidying up his premises, both in and outside, making his surroundings more healthful, more cheerful, and more beautiful, thus bringing a joy and comfort into his home that he has not known heretofore, and at practically no expense."[4]

To the end of his life, Carver continued to have a profound appreciation for the beauty inherent in creation, and his limited leisure time was often devoted to exploring the artistic potential of natural materials. In 1942, a writer from *Time* magazine, describing a visit to the new Carver Museum at Tuskegee, mentioned "embroideries on burlap, ornaments made of chicken feathers, seed and colored peanut necklaces, woven textiles...which the almost incredibly versatile Carver had turned out between scientific experiment and paintings."[5]

A few years earlier, Carver had invited a small group of guests to his living quarters at Tuskegee. One of them later wrote how he had showed them one of his paintings, a depiction of roses so impressive that they gasped when they saw it. The canvas, Carver explained, was made of a cornstalk material, but the source of the paint was even more unusual: "Workmen were excavating to put a new pipe under my steps," he explained, "and I used some of the clay they dug to create these colors. When you came into my 'Den' just now you were literally 'walking on roses.'"[6]

Study Questions
Chapter 5

As a young man, Carver gave up his art studies to take up science. But years later, he could still say, "I am an artist by taste, training, and profession...."

What continuing role did art play in Carver's life?

How did his artistic outlook influence his work as a scientist and teacher?

Carver showed poor people how to beautify humble homes and yards using plants and natural materials that lay close at hand.

Why did he think this important?

In what ways does natural beauty nourish people?

How could Carver's idea of using locally available plants and materials be used to improve the appearance of your community?

Carver's friend Glenn Clark wrote, "I never saw anyone love flowers as Dr. Carver does."

Why do you think Carver loved flowers so intensely?

How did the flowers he wore in his lapel contrast with his coat?

What did this contrast suggest about Carver's priorities in life? What do you think about those priorities?

"TWO OF THE GREATEST PRODUCTS THAT GOD HAS EVER GIVEN US"

Although Booker T. Washington acknowledged Carver's gifts as a teacher and researcher, he found him deficient as an administrator. Gradually, Carver's administrative duties were reduced, and devastating though this was to his ego, it allowed him to devote more attention to research and thus set the stage for his eventual fame.

A few years after Washington's death in 1915, Carver managed to limit his instructional duties to the Tuskegee summer sessions for teachers. This, too, left more time for the laboratory.[1] Increasingly, his work there was centering on an ironic realization: by helping farmers to diversify and become more productive, he had paved the way to unmarketable surpluses. In 1921, he recalled a conversation he had held in Alabama a few years before with a woman who was a large landowner:

> She said to me, "Carver, what are we going to do for a money crop? The boll weevil has

47

come....We are not going to be able to raise cotton profitably. What are we going to do? I understand that you advocate the raising of peanuts, but what are we going to do with them after we raise them?"

She had propounded a very hard question for me. I had been talking about peanuts and many of the farmers had gone into the peanut business. I had increased production but not increased consumption....On the way home, I said, "That is the question and something must be done for I really don't know."[2]

More than two decades after Carver's death, Clarence Mason, then director of the Carver Research Foundation at Tuskegee, commented on the significance of Carver's decision to shift his focus:

Now here was a man, a scientist, a mycologist whose primary interest was botany, who was a trained observer of plant life...who turned away from that primary interest to work in chemistry. By this time he was no longer a young man, and this transition must have been difficult for him, but he felt that his contribution to mankind would be more beneficial if he demonstrated the value of peanuts and sweet potatoes in industry....As a scientist I have the most profound respect and admiration for Dr. Carver because he turned away from the field in which he was most skilled and best trained, to work in an area which he felt he could do the most good for the people in his community.[3]

Carver, who came to believe that the peanut and sweet potato were "two of the greatest products that God has ever given us," devoted more and more attention to industrial uses for those two crops as well as food applications that went beyond simple home recipes. He developed a rubber compound from sweet potatoes, and during World War I, federal authorities concerned about wheat shortages consulted with him on his process for preparing a sweet potato flour that could be used in baked goods.[4]

In 1919, Carver dashed off an excited note to Robert Moton, Washington's successor as Tuskegee principal. In the note, Carver explained that he had just produced "a delicious and wholesome milk from peanuts." Carrying his experiments further, he produced cheese, buttermilk, ice cream, and other substitute dairy products from the same source.[5]

In 1920, Carver appeared before a convention of the United Peanut Association of America and made such an impression with his product samples and his engaging presence that the organization asked him to appear the next year before members of the Ways and Means Committee of the U.S. House of Representatives as they considered a tariff bill with a potential economic impact on the peanut industry.

Again, Carver was a hit, and this time the impression he made was a major milestone on his way to national fame. Initially granted ten minutes to speak, he so fascinated the congressmen that the committee chairman declared his time to be unlimited. He displayed not only peanut "dairy" products, but also breakfast food, instant coffee, Worcestershire sauce, face cream, ink, mock oysters, and other products—each made in whole or in part from peanuts or peanut byproducts![6]

Several years later, agricultural journalist Wheeler McMillen visited Tuskegee and asked Carver how he had gone about his exploration of the peanut and its uses. The reply, as reported by the journalist in a subsequent magazine article, was both simple and startling:

> **Why, I just took a handful of peanuts and looked at them. "Great Creator," I said, "*why did you make the peanut? Why?*"**
>
> **With such knowledge as I had of chemistry and physics I set to work to take the peanut apart. I separated the water, the fats, the oils, the gums, the resins, sugars, starches, pectoses, pentoses, pentosans, legumen, lysin, the ameno and amedo acids. There! I had the parts of the peanut all spread out before me. Then I merely went on to try different combinations of those parts, under different conditions of temperature, pressure, and so forth.**
>
> **The result was what you see—these 202 different products, all made from peanuts![7]**

Study Questions
Chapter 6

Although not impressed with Carver's skills as an administrator, Booker T. Washington saw that he had other gifts.

What were some of Carver's most notable gifts?

What gift was he finally able to emphasize, and how did this emphasis change his life?

Although Carver's principal training was in botany and mycology (the study of fungi), he turned to chemistry to find new uses for plants—particularly peanuts and sweet potatoes.

Why did he decide to do this?

What personal qualities does it take to make such a major transition in one's life?

Despite his high-pitched voice and despite the racial prejudice so prevalent in the 1920s, Carver was an immense hit when he spoke before a congressional committee.

How was he able to so impress the members of the committee?

What skills does Carver's success imply?

How did he develop these skills?

Early Years at
Tuskegee Institute

Later portrait,
date unknown

This painting won honorable mention at at the World's Fair in Chicago in 1893.

George Washington Carver displaying some of his paintings, including the award-winning work pictured above.

Multi-gifted as he
was, he enjoyed
painting one of his
favorite subjects...

...or knitting in his study.

George Washington Carver (third from left in front of window) with one of his classes at the Farmer's Institute at Tuskegee around 1910.

Dr. Carver exhibits some of the many products he developed from the peanut plant.

George Washington Carver
and a new friend at
Tuskegee Institute.

Still studying in
his later years.

George Washington Carver gave the Commencement
Address at Simpson College in 1942.

In the Word, early in the morning, "...all my life I have
risen regularly at four o' clock and have gone into the
woods and talked with God. There he gives me my orders
for the day."
GWC, quoted in Glenn Clark, *The Man Who Talks With The Flowers*

Statue at the George Washington Carver National
Monument, Diamond, Missouri

George Washington Carver and friends at
the Booker T. Washington memorial at Tuskegee

Gathering the "fruits of creation..."

...for God's discovery process, in his lab

"GOD SPEAKING TO MAN THROUGH THE THINGS HE HAS CREATED"

The novelty value of Carver's research, combined with his compelling personality and humble origins, made him a journalist's dream. Some writers, unfortunately, embellished what would have been a fascinating biography left unembellished. As the list of Carver's products grew, so did the popular mythology that portrayed him, in seemingly contradictory terms, as a great scientist whose work bordered on wizardry.

With his deep Christian faith and mystical nature, Carver himself added to this confused image. He told of arising each morning at four to walk in the woods, commune with the Creator, and receive his orders for the day. He called his laboratory "God's Little Workshop,"[1] and he liked to quote scriptural passages that he believed had particular relevance to his work and thought.

In an article published in a Baptist periodical, a writer told how he asked, "What, Dr. Carver, is the most marvelous fact of the age, or of the ages, the most wonderful conception of your mind?" Carver's answer, said the writer, was "immediate and like a flash":

> **The creation story, the creation of the world. "In the beginning God...created the heavens and the earth...and God said, Behold, I have given you every herb yielding seed, which is upon the face of all the earth, and every tree, in which is the fruit of a tree yielding seed; to you it shall be for food:..."**
>
> **"Behold" there means "look," "search," "find out"...That to me is the most wonderful thing of life.**[2]

In an essay published well before his rise to national fame, Carver described how beholding creation could bring a closer relationship with the Creator:

> **To me Nature in its varied forms is the little windows through which God permits me to commune with Him, and to see much of His glory, by simply lifting the curtain and looking in.**
>
> **I love to think of Nature as wireless telegraph stations through which God speaks to us every day, every hour, and every moment of our lives.**[3]

In 1924, Carver addressed an audience at Marble Collegiate Church in New York. As reported in an Associated Press story, he asserted that "No books ever go into my laboratory. I never have to grope for methods; the method is revealed at the moment I am inspired to create something new. Without God to draw aside the curtain, I would be helpless."

An anonymous editorial writer in the New York *Times* found such talk deplorable. "Real chemists, or at any rate, other real chemists," said the writer, "do not scorn books out of which they can learn what other chemists have done, and they do not ascribe their successes, when they have any, to 'inspiration.' Talk of that sort simply will bring ridicule on an admirable institution and on the race for which it has done and still is doing so much."[4]

Carver believed his words had been seriously misinterpreted. In a response to the editorial he wrote,

> **I regret exceedingly that such a gross misunderstanding should arise as to what was meant by "Divine inspiration." Inspiration is never at variance with information; in fact, the more information one has the greater will be the inspiration.**

He went on to name many chemists who had influenced him and noted that he received "the leading scientific publications."[5]

In another response to the *Times* editorial, Carver expanded on his definition of inspiration. A writer for the *Golden Age*, a Jehovah's Witness magazine, quoted him as follows:

> **I know that my Redeemer liveth. I know the source from whence my help comes. Inspiration, as I used the word in my New York lecture, means simply God speaking to man through the things He has created; permitting him to interpret correctly the purposes the Creator had in permitting them to come into existence. I am not interested in any science that leaves God out; in fact, I am not interested in anything that leaves out God.[6]**

To the end of his life, Carver's approach to his work closely combined his profound appreciation for creation with his faith in the Creator. Recalling Carver's visit in 1939 to the Starr Commonwealth home for boys in Albion, Michigan, Christy Borth told how Carver had held a youthful audience spellbound for three hours:

> **What began as a lecture on botany soon developed into a soul–stirring recital of how intimately all of the plants are related to one another, of how the plants and the animals—mankind included—are inextricably interdependent, and of how the whole of creation is related to its Creator.[7]**

Perhaps Carver's approach to life is best captured in a story related by his Tuskegee colleague, B.B. Walcott. Describing the opening of the Carver art collection at the school in 1941, the writer told of the aging educator's response to someone who asked how he had done so many different things in his life:

"Would it surprise you," he replied gently, "if I say that I have not been doing many DIF-FERENT things?...All these years...I have been doing one thing. The poet Tennyson was working at the same job. This is the way he expresses it:

Flower in the crannied wall,

I pluck you out of the crannies,

I hold you here, root and all, in my hand,

Little flower—but if I could understand

What you are, root and all, and all in all,

I should know what God and man is.

"Tennyson was seeking Truth. That is what the scientist is seeking. That is what the artist is seeking....My paintings are my soul's expression of its yearnings and questions in its desire to understand the work of the Great Creator."[8]

Study Questions
Chapter 7

In a lecture at the Marble Collegiate Church, Carver told of receiving God's help while working in the laboratory.

Why was Carver criticized for his comments?

How did he define the kind of inspiration he received?

Have you ever experienced a similar kind of inspiration in your life?

Christy Borth recalled a lecture in which Carver described "how the plants and the animals—mankind included—are inextricably interdependent, and...how the whole of creation is related to its Creator."

What did Carver mean when he said plants, animals, and humans were interdependent?

Can you think of an example of this interdependence?

How is the "whole of creation" related to the Creator?

Carver said that the different things he had been doing for many years were really just one thing.

What was the "one thing" Carver had ultimately been trying to do?

How well did he succeed in his quest?

"THE FIRST AND GREATEST CHEMURGIST"

C hemurgy is now an almost forgotten word, yet it once identified one of the most progressive and farsighted movements in American history—a movement that George Washington Carver anticipated by two decades.

The word itself, meaning "chemistry at work," was coined by noted organic chemist William J. Hale who, together with Wheeler McMillen, began in the late 1920s to promote the idea that farm crops and crop residues (such as corn stalks and oat hulls) should be used more extensively as raw materials in industry.

In 1935, Hale and McMillen, joined by automaker Henry Ford and a number of prominent figures in agriculture, science, and industry, met for a historic conference at Dearborn, Michigan. Out of that conference grew the Farm Chemurgic Council. The Council's program, which initially placed considerable emphasis on promoting the use of ethanol, made from farm crops, as a supplement to gasoline, received much favorable attention in the press. The country was in an economic depression, and the chemurgists believed that fuel and consumer goods made from biomass were important keys to helping hard–pressed farmers market their surpluses. Furthermore, a number of other countries were already using gasoline/alcohol blends, and concerns were being voiced in some quarters that America's oil reserves would be depleted before many more years had passed. In this atmosphere, the idea of an American renewable–fuel industry had a certain resonance.[1]

The chemurgists were solar pioneers, and their emphasis on renewable energy and renewable raw materials sound all the more prophetic in a time when national leaders are willing to spend billions of dollars and place countless lives on the line to protect supplies of imported petroleum. But these pioneers recognized that someone else had been far ahead of them. Following his visit to Tuskegee in 1928, McMillen had reported in a national magazine article that Carver's peanut products by then included seventeen different wood stains, shoe and leather blacking, axle grease, cloth dyes, linoleum, pomade, antiseptic soap, and shampoo. Furthermore, the Tuskegee researcher had incorporated sweet potatoes into library paste, ink, dyes, and shoe blacking; produced wallboard from elephant ear; and made paper products from mulberry, chinaberry, yucca, and palmetto.

Little wonder that journalist Christy Borth, in a 1939 book chronicling the chemurgy movement, would see fit to dub Carver the "first and greatest chemurgist."[2]

The same year Wheeler McMillen visited Tuskegee, he interviewed Henry Ford, who also expressed interest in renewable raw materials for industry. This mutual interest in chemurgy would later bring together the industrial titan and the black professor in one of the unlikeliest friendships of the early 20th century.

The two men first met in 1937 when Carver went to Michigan to speak at a national chemurgy conference. They quickly became mutual admirers, sharing an interest in natural foods as well as renewable raw materials. Although both men were in their seventies by the time they met, Carver's letters to his "great inspiring friend," were filled with the kind of enthusiasm one might expect of a boy who discovers that a classmate shares his hobbies.[3]

Carver and Ford both deplored waste. Carver was greatly impressed when his friend, visiting Tuskegee after a stay at his winter estate in Ways, Georgia, ordered his chauffeur to recover a bottle that had fallen out of his car:

> **[Ford] owns about seventy–five thousand acres of land in Ways....He had torn down an old cabin on this land and had picked up this bottle underneath that house—and he said that it was**

valuable to him, we can use it. He took it way back to Michigan. I looked at it and said, "Well, well, there is the richest man in the world picking up an old dirty bottle to carry back to Michigan...."[4]

Ford had grown up on a farm, and like Carver, his concern for the welfare of farmers inspired some of his most farsighted ideas. He believed that he could grow, rather than mine, more of the raw materials for his automobiles, and strange though that idea may seem in our day, he demonstrated that it was true.

Carver, who lacked a business orientation, never saw more than a handful of his products enter the marketplace, but Ford was able to show quite convincingly that his favorite plant, the soybean, had versatile industrial potential. His team of youthful researchers developed a soy–based exterior enamel for Ford automobiles, and soy meal was incorporated into such parts as the gearshift knob and horn button. The Ford researchers also developed a fabric from soybeans, and in 1941, they completed a prototype car with a plastic body derived largely from soybeans and other farm–grown ingredients.

Ford, with his immense wealth and huge corporate infrastructure, was also able to go farther than Carver in demonstrating the potential for regional self–reliance based on local resources. He employed part–time farmers at small Ford "village industries" in Saline and Milan,

Michigan, where soybeans from the surrounding region were processed for industrial uses. Drawing some of their power from small hydroelectric plants, the two facilities used local renewable energy, processed local renewable raw materials, and benefitted local communities by providing additional income to local farmers.[5]

Ford's Saline and Milan operations were excellent examples of the emphasis on local communities and regions implicit in chemurgy. Agricultural commodities were bulky, so it made sense to process them for industry close to the point of harvest. The Ford factories were also visible reminders of Carver's long–time contention that people could learn to appreciate and make better use of locally available renewable resources. As early as 1902, he had commented in a magazine article on the senselessness of a breakfast he had recently eaten in his own state of Alabama: bacon from Kansas, grits from Massachusetts, flour from Nebraska, oranges from Florida, bananas from Cuba, sugar from Louisiana, and coffee from Java. Only the milk and butter, he observed, had come from nearby.[6]

Carver stressed that perfectly satisfactory alternatives could be grown—or already grew wild—at home. In a story related by his biographer Rackham Holt, Carver gave a tramp some money to buy food, but commented to a companion on the absurdity of the situation. After all, he said, along the route to the store were edible plants sufficient to feed a whole town a balanced diet.[7]

In 1927, Carver visited Tulsa, Oklahoma. In a speech delivered there, he told of getting up that morning, taking a walk up nearby Stand Pipe Hill, and discovering numerous plants indigenous to Oklahoma that had medicinal properties. Later, he said, he had visited a local drug store and found "seven patent medicines containing in their formulas certain elements contained in these plants on Stand Pipe Hill. The preparations were shipped in from New York. They should be shipped in from Stand Pipe Hill."[8]

But for Carver, developing the kind of "creative mind" needed to fully recognize the resources that lay beneath one's feet was apparently only a hopeful beginning. After returning from a trip to Kansas, he wrote enthusiastically to a man in Wichita about the state's rich endowment of clays, vegetable dyestuffs, and medicinal plants. He then proceeded to paint a remarkably edenic word picture of a society working harmoniously to develop the fruits of creation:

> **When one fully realizes that every farm, garden and orchard product will yield new, strange, and useful things to the thoroughly developed creative mind, an inexhaustible [sic] field of possibilities dawns upon us; a field in which all can work without clashing, indeed the greater the number of workers, the more interesting their work becomes, and the more closely they are drawn together, as here we really walk and talk with the Great Creator; it is here that He shows**

His glory, majesty and power in such an understanding and unmistakable way.

It is in this realm that our minds are lifted above the sordid things of life. We have no room for a religion of hate, because God expresses Himself in everything He has created, hence we love it, because God is love, and man the highest embodiment of His handiwork, naturally, we should love most.[9]

Study Questions
Chapter 8

Renewable resources are those that nature's processes can replenish and sustain. Examples might include fuel or plastics made from plants.

Why did Carver and his friend Henry Ford place such emphasis on renewable resources?

Generally, what are the advantages of using renewable rather than non-renewable resources?

Do you think renewable resources will play a more important role in our future? If so, why?

Can you think of ways in which renewable resources have been misused? How can we assure that they are replenished and sustained?

Carver deplored waste and appreciated the fact that his friend Henry Ford went out of his way to save an old bottle that might have some future use.

Name some of the things that we waste or throw away which might have some future use.

What does wasting things say about our attitude toward God's creation?

What effects does wastefulness have on our communities? On the environment?

Do churches sometimes engage in wasteful behavior? What can they do to set a good example for their members and the surrounding community?

Carver described a breakfast he had that included many ingredients which were not locally grown. Why did he think such a breakfast did not make sense?

Where do your breakfast foods come from, and what resources and labor are required to bring them to your table?

What would be the value of eating locally grown foods?

If you were to eat a breakfast comprised mainly of foods grown in your own state or region, what would it include?

"NATURE'S REMEDIES, WHICH GOD INTENDED WE SHOULD USE"

In 1916, a writer described how Carver himself used plants for "medicinal purposes." He would eat

> tomatoes for this, beans for that, rape for another trouble, cabbage for another, watercress for another, liquor of pine needles for colds, dandelions for something else. He knows and eats a score of vegetables that other people sneer at as weeds. He has a small range in his room, and when the bill of fare in the dining room is not to his liking or to the benefit of his health he goes out into the seemingly barren fields, brings in things (I have no better word), cooks and eats them, and is happy and healthy.

Two years earlier, Carver had written to Booker T. Washington that "there is probably no subject more important than the study of foods in relation to their nutrition and health." Noting that he had not been sick in bed for "quite 35 years," he explained to Washington that he regulated himself "with vegetables, fruits, and wild herbs, Nature's remedies, which God intended we should use."[1]

In one of his Experiment Station bulletins, *Three Delicious Meals Every Day for the Farmer*, Carver said, "Fresh fruits and vegetables have a medicinal value, and when wisely prepared and eaten every day will go a long way towards keeping us strong, vigorous, happy, and healthy, which means greater efficiency and the prolonging of our lives." He went on to emphasize the economic impacts of illness, pointing out that a person with a small salary could "save but little if anything when someone in the family is sick almost constantly." On the other hand, for readers who carried out his suggestions for a balanced diet, he promised both tangible and intangible benefits: "[Y]ou will be surprised how much healthier, happier, and how much more work you can do; and how quickly you will become self–supporting."[2]

Carver also saw a connection between poor nutrition and crime. Speaking before a class in dietetics at Tuskegee, he explained his theory:

> **It may seem startling when I say that the majority of our criminals are produced as a result of bad cooking and bad combination of food stuffs. But upon serious thought, one can see that this is true. When a person eats poorly cooked food, his body is not properly nourished. A poorly nourished body produces an unhealthy mind. A person who is poorly nourished quickly turns to stimulants, such as alcohol, to give him a feeling of well being, which should have been supplied by properly cooked food.[3]**

Chapter Nine

In 1929, Charles Freer Andrews, a close associate of Mahatma Gandhi, visited Tuskegee. During the visit, Carver presented him with a diet for the ascetic leader of India's anti–colonial struggle, including a milk substitute based on soybeans. Afterward, he wrote Andrews as follows:

I believe Mr. Gandy's [sic] physical strength can be greatly improved by following out the ideas we discussed in the matter of foods.

With the whole wheat flour, grits, hominy, graham flour, etc., etc., which can be made on the little mill shown you, and with the splendid native fruits and vegetables you have, properly compounded, will give you a splendid nourishing and palatable food stuff.

You can use it in your school, they will in turn carry the message into the various communities from whence they came, bringing to my mind greater health, strength and economic independence to India.[4]

Unlike Gandhi, Carver was not a vegetarian, but he was interested in plant–based substitutes for meat as well as milk. Exhibiting mock oysters made from peanuts during his 1921 congressional testimony, he said, "We are going to use less and less meat just as soon as science touches these various vegetable products and teaches us how to use them."

In part, this interest in substitute foods reflected his concern for poor people, who could not always obtain animal protein. For example, he provided a recipe for soybean milk to missionaries in the Belgian Congo struggling to save the lives of undernourished babies, and his bulletin *How to Grow the Peanut and 105 Ways of Preparing It for Human Consumption*, largely aimed at poor farm families, included recipes for mock chicken, mock veal cutlets, and mock sausage.[5]

Carver also shared Henry Ford's notion that the cow was a crude milk producer. Both men believed it made sense, from the standpoint of efficiency and sanitation, to simply bypass the cow, producing "milk" directly from the plant world. Ford sponsored extensive research on soybean "dairy" products, much of it conducted in a nutritional laboratory at Dearborn, Michigan, which was named for Carver. Although the laboratory, established in 1942, remained in operation only four years, its work inspired several innovative people who went on to manufacture such now–familiar products as non–dairy coffee creamers and whipped dessert toppings.[6]

These products hardly fit with Carver's hopes, expressed in a letter to Ford, that the laboratory would help people learn "how to live as nature intended it,"[7] but their ubiquitous presence in supermarkets did raise public awareness that dairy substitutes were possible. Today, "milks" based on soybeans, rice, almonds, or even potatoes are staples in the natural–foods industry, as are nutritious plant–based substitutes for cheese, ice cream, hamburgers, and hot dogs. As world population explodes, land becomes more scarce, and the ecological costs of large–scale livestock agriculture continue to mount,[8] such products, made from locally available grains and vegetables, may well grow in importance. Perhaps in regions where the peanut thrives, people will enjoy "Carver milkshakes" with "Carver burgers" and weed–based "Carver salads."

Study Questions
Chapter 9

As a child Carver was frail and sickly. Yet when he was well into middle age, he could look back and write that he had not been sick in bed for "quite 35 years." He explained that he made use of "Nature's remedies": vegetables, fruits, and wild herbs.

What do you believe to be a healthful diet?

Can you think of any herbal remedies that are commonly used today?

Do you see a relationship between being a good steward of your own body and a good steward of creation?

Carver tied poor nutrition to crime. He said "a poorly nourished body produces an unhealthy mind," and "a person who is poorly nourished quickly turns to stimulants, such as alcohol...."

Do you think Carver was right?

How much do you think a person's mood and behavior can be influenced by what he or she eats?

Carver recognized that poor people could not always obtain meat and dairy products. He sought to provide nutritious alternatives made from peanuts and other plants.

Can you think of other reasons why such food alternatives might be important in today's world?

While people in other parts of the world struggle to get enough grain to feed their families, Americans feed large quantities to livestock raised for meat production. Do you see anything wrong with this?

"WE HAVE EVERYTHING WE NEED"

As the chemurgy movement began to blossom in the mid 1930s, recognition of Carver as a prophet grew. He was energized by his new friendship with Henry Ford and pleased by his association with Austin W. Curtis Jr., a young Cornell graduate who became his research assistant in 1935. Working with Carver, Curtis did research in such areas as dehydration of sweet potatoes and development of low–cost paints made from local clays and used motor oil.[1] For the never–married Carver, Curtis became a surrogate son and indispensable aide.

With new attention focused by the chemurgists on ideas that Carver had espoused for decades, it suddenly seemed possible that America would begin to move toward an economy based more closely on local, renewable resources. With the chemurgists paying particular heed to the potential of plant–based motor fuel, Carver pointed to a variety of potential alcohol feedstocks in the South, ranging from sweet potatoes to over–ripe oranges. He expressed his hope that "the production of ethyl alcohol from home–grown products and waste will be our next successful venture."[2]

Nevertheless, despite the interest of journalists and the general public, American industry continued to ignore Carver's own chemurgic wizardry. Shortly before America's entry into World War II, a reporter visited Carver at Tuskegee and noted his frustration:

> **The aging scientist waves a hand toward the exhibit cases. His high pitched voice is earnest. "We've given it to them," he says, referring obviously to those who could bridge the gap between laboratory and the consumer. "It's up to them to use it, make it available to the people. It's pitiful, pitiful the way they're letting things go to waste."**

But Carver expressed his hope to the same reporter that the war then spreading around the world might hold some positive lessons for Americans, "teaching us to utilize what we have, not depending so much on imports. We have everything we need, if we would only use it."[3]

World War II did bring a last respectful look at Carver's ideas about better utilization of renewable resources. With most of the nation's vital supplies of rubber cut off by the Japanese, rumors circulated in 1942 that Carver, in cooperation with Henry Ford, would come up with a rubber substitute. The rumors came to nothing, and Carver admitted to the federal War

84

Production Board that his work on producing rubber from sweet potatoes had been abandoned many years earlier, without progressing into "the essential pilot plant or process development stage that it should have." However, he emphasized to the press that there were many other plant sources of rubber. Indeed there were, and during the war the government experimented with some of them, including guayule, a southwestern shrub, and kok–saghyz, otherwise known as Russian dandelion.[4]

The government also sponsored other programs to make better use of domestic renewable materials. School children were encouraged to gather milkweed floss, which served as a filler for military life preservers in place of kapok, formerly imported from southeast Asia. And farmers under government contract grew hemp (previously condemned under its other name, marijuana, as a dangerous drug source) to provide raw material for rope and cordage.[5]

Prospects for renewable raw materials in industry were bright into the 1950s, but as time went by, a burgeoning petrochemical industry pushed chemurgy into the background. Today, however, environmental concerns are helping to spark new interest in renewable fuels and feedstocks. Renewable, biodegradable plastics are entering the market, and soybean inks, which emit fewer polluting volatile organic compounds (VOCs) than their petroleum–based counterparts, are rapidly gaining popularity. Carver, with his concern for transforming wastes into resources, would take special delight in current research interest in transforming municipal solid waste into alcohol fuels.[6]

Recently, farsighted leaders in science, industry, and agriculture have joined together in a new effort to promote such developments. In March 1990, more than six decades after visiting Carver at Tuskegee, 97–year–old Wheeler McMillen addressed a national conference in Washington, D.C., which served to launch this latter-day chemurgy movement. In words echoing Carver's own hopes, he predicted the conference would "inspire work for years to come, bringing more stability and prosperity to farmers, and assuring a better environment for all of us from the Earth's bounty."[7]

Study Questions
Chapter 10

In the 1930s and '40s, it seemed that Carver's hopes for the future of chemurgy—the branch of chemistry devoted to making non-food products from plants or trees—were well-founded.

Why was chemurgy important during World War II?

Why did it become less popular later?

Why have some leaders in science, industry, and agriculture taken a new interest in chemurgy in recent years?

Carver was very adept at finding new uses for plants. He saw potential that other observers had overlooked.

Did his ability to see potential derive more from his scientific skill or his attitude toward creation and the Creator?

Have you ever found unexpected potential in creation (e.g., a beneficial use for a plant normally considered to be a weed)?

Do you think Carver's openness to unexpected potential had something to do with the fact that he belonged to a race whose potential was long denied due to prejudice?

Was Carver concerned only with the usefulness of plants for people, or did he also see them as having intrinsic value?

In 1941, shortly before America entered World War II, Carver was quoted as saying the United States had all the resources it needed.

Do you think he would be able to make this statement if he were alive today?

Should we stop importing so much petroleum and place more emphasis, as Carver did, on the potential for producing fuels and other products from renewable resources available in our own country?

Would it be easier for the United States to maintain peaceful relations with other countries if it was not so dependent on imported resources, such as petroleum?

"I AM A BLAZER OF TRAILS"

On January 5, 1943, George Washington Carver died, and the nation, preoccupied with news of the war, nevertheless took time to mourn the orphan son of slave parents who had risen to national fame.

In her admiring popular biography, published only months after Carver's death, Rackham Holt made an astute observation about her subject's legacy:

> His discoveries, with the exception of his mycological work, did not properly belong in scientific journals. They were not revolutionary in themselves. Anyone with the proper education could milk the peanut or abstract paper from suitable fibers, or rubber from the sweet potato or any other vine which secreted latex. His special contribution was to expose these hidden properties in plants to the public view and, by dramatizing them, serve as a signpost pointing the way for those who had the facilities to incorporate them into the contemporary pattern of living.

"I am a blazer of trails, new trails," Carver himself had said. "Little of my work is in books. Others must take up the various trails of truth and carry them on."[1]

In the decades following his death, few chose to ask where his trails might lead. A consumer society worshiping the temporary gods of abundant petroleum and cheap petrochemicals had little appreciation for someone who advocated renewable fuels and developed myriad products based on plants. A nation building its prosperity on a mountain of waste had no time for a prophet who deplored the very concept of waste. And a mobile, urbanized population could not relate to a rural agriculturist who showed how people could remain in the countryside, enriching their lives and restoring their land through better understanding of nature's gifts.

Quoting from the book of Proverbs, Carver would remind his listeners that "Where there is no vision, the people perish." Today's global environmental dilemma underlines the crucial importance of that reminder, and in recent years, thoughtful observers have come to recognize what Carver knew instinctively: that people and their governments cannot hope to triumph over poverty, hunger, and disease without also addressing such underlying threats to the human home as deforestation and soil erosion.

It is this recognition that led to the historic 1992 Earth Summit in Rio de Janeiro. Despite the political wrangling that prevented the Summit from fully addressing some crucial issues, historians of the next century may well look back on the Rio gathering as a watershed event in both human and ecological history. The prescriptions of Agenda 21, the Summit's major policy document, may prove to be the hopeful starting point for a whole new way of ordering life on earth: one that respects both human needs and the long–term welfare of the planet.

Although Carver was a pioneer in recognizing the connection between human and ecological concerns, many Earth Summit issues were either unrecognized or of lesser magnitude in his lifetime. Among these we may count population growth, ozone depletion, global warming, destruction of tropical forests, and accelerating species extinctions. Furthermore, as a basically apolitical man working in an institution guided by the cautious philosophy of Booker T. Washington, Carver only indirectly confronted such persistent international development issues as inequitable land distribution and the power of landlords. For these reasons, many of today's environmentalists and development specialists would be inclined to dismiss him as a historical curiosity.

And they might even fail to recognize him as one of their own. Far from resembling an environmentalist as that term is currently understood, Carver was a potentially puzzling combination of nature mystic, saint, scientist, and business booster. He certainly stood in awe of creation as he found it, but his wonder was combined with a sense that God had placed in nature vast potential for human betterment. He was a St. Francis armed with test tubes, seeking, through scientific means, creation's undiscovered fruits to enhance the well–being of all people.

Although he fully recognized that people were part of a web of natural relationships with a profound responsibility to understand and work harmoniously within that web, Carver stood in sharp contrast to some current environmentalists by placing the human species above all others as the "highest embodiment" of God's handiwork. And, while his attempts to market his own products faltered, he was a notable business promoter in his role as unofficial spokesman for the American peanut industry.

But on reflection, it is clear that Carver, in his own complex personality, anticipated a broadening of the environmental constituency that would have seemed impossible even a decade ago. Christians who want to "till and keep" the earth in ways consistent with their faith are joining the environmental ranks in increasing numbers. More scientists, distressed over bleak environmental prospects, such as global warming, are raising their voices in alarm. And progressive members of the business community are coming to recognize that in the face of mounting pollution and diminishing resources, industry must respond with leaner waste streams, cleaner production processes, and "greener" consumer goods.

To these groups, and to the many others that must join in tackling the challenges outlined in Agenda 21, Carver offered an intriguing vision. It was centered on the American South of his day, but adaptable, to a greater or lesser degree, to other regions of the world in our day. In short, Carver saw a self–reliant society, carefully nurturing its land and creatively using its resources—including resources unjustly classed as wastes or weeds—to enhance the health and security of all its residents.

This vision was rooted in his belief that even the humblest of God's creations has a purpose. In fact, he saw the very concept of waste as illusory. A century ago, he wrote:

> **The earnest student has already learned that nature does not expend its forces upon waste material, but that each created thing is an indispensable factor of the great whole, and one in which no other factor will fit exactly as well.**[2]

In a sense, Carver's entire career centered on demonstrating that everything in creation has a purpose. A Tuskegee colleague observed that as far as Carver was concerned, "nothing was thrown away": he found many uses for scraps of paper, bits of string, or stubs of pencils. His first

Tuskegee Experiment Station bulletin was aimed at encouraging better use of acorns, which had been "hitherto practically a waste product." One of his last projects was production of a soap with "perfectly marvelous" lathering properties, made in part with refuse scraped from the floor of a peanut–shelling plant.[3] Confronted with the immense wastes generated by today's industrial societies, Carver would bemoan the blindness that led to such absurdity, then quickly focus on discerning each waste product's real purpose as "an indispensable factor of the great whole."

And he would surely be at the forefront of current efforts to preserve the earth's threatened biodiversity, condemning the wasteful destruction of tropical forests and pointing to the great numbers of tropical species still unexamined for their potential as sustainably harvested sources of foods, medicines, essential oils, fuels, fibers, gums, and adhesives.[4]

If everything in creation has a purpose, it should be possible for the rural poor, by recognizing and fully employing the resources at their doorstep, to live more abundantly without expending scarce capital. Much of Carver's attention was devoted to demonstrating how the poorest farm families in his vicinity could restore their degraded land, diversify their diets, improve their health, and beautify their surroundings, all at little or no cost. It simply meant recognizing and using creatively the resources already at hand.

In one sense, this was thoroughly traditional. Like Native Americans and members of other indigenous cultures, Carver was an intuitive naturalist, sensitive to his locality's broad potential for human sustenance. But he combined his intuitive wisdom (and, he would say, the Creator's inspiration) with the tools and techniques of modern science. As a scientist, he was able to discover potential in the natural world that would be hidden to traditional peoples. As an intuitive naturalist, he learned things not readily apparent to an industrial society that sought to overpower, rather than understand, nature.

Carver's combined approach made possible the kinds of insights achieved by today's ethnobotanists as they work to discover pharmaceutical sources in collaboration with native rain–forest plant–lore specialists.[5] We can hope that such collaborations with local people, in the rain forests and elsewhere, will prove to be a key element in preserving biodiversity by pointing to new economic possibilities for sustainably harvested plants.

When Carver shifted from a focus on family self–reliance to one of regional prosperity, he was again blazing a trail to a sustainable future. He dreamed of the day when his region would produce renewable fuels and a host of other products from under–utilized plants and waste products. It is true that the American South in his time, despite human–induced ecological degradation, had natural advantages not shared by every region. Nevertheless, Carver, with his focus on purpose and possibilities, surely would have found intriguing new economic prospects—based on wastes, weeds, and other previously overlooked local resources—almost anywhere on earth.

Today, as desperate rural people around the world crowd into polluted cities, hoping to somehow better their lives, Carver's vision raises an intriguing question: How many of those people could remain where they are, preserving the best of their traditional cultures while living in harmony with their environment and launching economic enterprises based on local resources? Certainly, developing countries must not emulate the mistakes of Europe and North American in building massive, polluting industries based on non–renewable fuels and petrochemical raw materials. Carver's vision, combined with the village industry model provided by his friend Henry Ford, points to a more sustainable alternative: smaller–scale local industries, efficiently using renewable energy and raw materials, recycling all wastes, and employing local people to produce lubricants, plastics, building materials, soaps, stains, dyes, paints, and a host of other useful products.

The challenge in an increasingly crowded

world will be to assure that renewable resources are used in an equitable, sustainable manner. In the early 1980s, knowledgeable observers began to consider whether expanding demand for ethanol made from farm crops, such as grains, could touch off a "food vs. fuel" competition between the world's wealthy automobile owners and its poorest residents, for whom obtaining sufficient food was already difficult. To such concerns, Carver would likely respond by emphasizing recent research aimed at making non–crop resources—such as forestry residues and municipal solid wastes—economically attractive feedstocks for alcohol fuel. He might also add that the world's poor could make better use of local weeds to supplement the grains in their diet, and that the world's wealthy could make more grain available by channeling less of it through beef cattle, which return in their meat only a fraction of the calories and protein that could be obtained by people eating the grain directly.[6]

For Carver, it would be a matter of reverently assessing the purpose, or set of purposes, for each element of God's creation and considering how that element related to others in the "great whole."

Carver's own profound appreciation of nature and his ability to "understand relations" may have stemmed initially from his genius, but he believed such qualities were basic and teachable. In fact, in his early days at Tuskegee, he played a major role in building an educational program for illiterate and half–educated people around that belief. Realizing his larger vision for nature and society will require a similar kind of educational outreach effort on a global scale—one designed to bring the workings of nature "down to the everyday life and language" of ordinary people and supply them with the knowledge of the "mutual relationship between the animal, mineral, and vegetable kingdoms, and how utterly impossible it is for one to exist in a highly organized state without the other."

Speaking of economic conditions in the southern United States of his day, Carver gave advice that is equally relevant to Agenda 21 goals in our day:

> **We should form ourselves into a committee of one—each one, if possible, feeling the burden more strongly than the other—and with concerted action, inaugurate a mighty campaign of education, which will lead the masses to be students of nature.[7]**

Study Questions
Chapter 11

Carver described himself as "a blazer of trails" and said, "Little of my work is in books."

What kind of trails did Carver blaze?

Where do his trails lead?

What are the lessons of his methods?

How can we take up his "trails of truth" and carry them on?

Carver reminded his listeners that "Where there is no vision, the people perish."

What was Carver's vision for society?

Did he set an example in his own life consistent with his vision?

What do you think his vision has to offer us as we struggle to resolve the difficult social and environmental issues facing today's world? What is it lacking?

Carver died in 1943 before the emergence of some of the important social and environmental issues that face us now. If he were alive today, what do you think he would say about the following:

Endangered plant and animal species.

Air and water pollution.

Crowded slums in many of the world's cities.

Heavy use of resources—both renewable and non-renewable—in the wealthier countries.

NOTES

Much of the research for this book was conducted in John W. Kitchens and Lynne B. Kitchens, eds., *The George Washington Papers in the Tuskegee Institute Archives* (Tuskegee, Alabama: Tuskegee Institute; Ann Arbor, Michigan: University Microfilms International [distributor], 1975). This 67–role microfilm publication includes Carver correspondence, correspondence of other people relating to Carver, articles by and about Carver, Carver memorabilia, and miscellaneous materials relating to Carver. Items from the microfilm publication are hereafter cited as "CM" (for Carver Microfilm) followed by the pertinent roll number (e.g., CM/43).

Other information was drawn from materials held by the Library of Congress (Washington , D.C.), the National Agricultural Library (Beltsville, Maryland), the Ford Archives (Dearborn, Michigan), the Soyfoods Center (Lafayette, California), and materials from the author's collection on the chemurgy movement.

The only previous scholarly biography of Carver (and by far, the best of the many biographies examined by this writer) is Linda O. McMurry, *George Washington Carver, Scientist and Symbol* (New York: Oxford University Press, 1981). McMurry's citations were very helpful in locating background material for this book. A very readable popular biography is Rackham Holt, *George Washington Carver, An American Biography* (Garden City, New York: Doubleday, Doran and Company, 1943), which was written with the cooperation of Carver and his assistant, Austin W. Curtis Jr.

CHAPTER 1: "MY VERY SOUL THIRSTED FOR AN EDUCATION"

1 George Washington Carver (hereafter cited as GWC), "Nature as Our Greatest Educator," *Guide to Nature*, Oct. 1908, 216; GWC, untitled biographical sketch, ca. 1897 (CM/1).

2 Anna Coxe Toogood, *Historic Resource Study and Administrative History, George Washington Carver National Monument, Diamond, Missouri* (Denver, Colorado: National Park Service, July 1973), 8–21 (CM/59).

3 GWC, untitled biographical sketch; GWC, "A Brief Sketch of My Life," ca. 1922 (CM/1).

4 For scholarly treatment of the kidnapping story and other details of Carver's childhood and

early adulthood, see McMurry, 9–28, and Robert P. Fuller and Merrill J. Mattes, "The Early Life of George Washington Carver" (Diamond Grove, Missouri: George Washington Carver National Monument, typescript, Nov. 26, 1957). (CM/59).

5 Recollections of Forbes Brown, May 20, 1952, in Fuller and Mattes, 27.

6 GWC, "A Brief Sketch of My Life."

7 Ibid.; McMurry, 18–24; Fuller and Mattes, 44–67.

8 GWC, untitled biographical sketch; GWC, "A Brief Sketch of My Life"; McMurry, 24–28.

9 McMurry, 31; Jessie L. Guzman, notes of interview with Etta M. Budd, June 16, 1948 (CM/63); GWC to L.H. Pammel, May 5, 1922 (CM/1).

10 Charles D. Reed to GWC, Dec. 22, 1932 (CM/13).

11 McMurry, 39–40.

CHAPTER 2: "TO BETTER THE CONDITIONS OF OUR PEOPLE"

1 For background on the life of Booker T. Washington and the early years of Tuskegee, see Louis R. Harlan's two–volume biography, *Booker T. Washington: The Making of a Black Leader, 1856–1901* (New York: Oxford University Press, 1972); and *Booker T. Washington: The Wizard of*

Tuskegee, 1901–1915 (New York: Oxford University Press, 1983)

2 McMurry, 44.

3 Thomas Monroe Campbell, "George Washington Carver – The Man," script for WAAF (Chicago) radio broadcast, July 11, 1948, Thomas Monroe Campbell Papers, Tuskegee University Archives; GWC, "What Chemurgy Means to My People," *Farm Chemurgic Journal*, Vol 1, No. 1 (Sept. 17, 1937), 40.

4 Max Bennett Thrasher, *Tuskegee: Its Story and Its Work* (Boston: Small, Maynard and Company, 1900), 104–105.

5 Thomas Monroe Campbell, *The Movable School Goes to the Negro Farmer* (New York: Arno Press and the New York *Times*, 1969 [first published in 1936]), 80–81.

6 Ibid., 68.

7 Booker T. Washington, "Twenty–five Years of Tuskegee," *World's Work*, Apr. 1906, 7441.

CHAPTER 3: "BEING KIND TO THE SOIL"

1 Allen W. Jones, "The Role of Tuskegee Institute in the Education of Black Farmers," *Journal of Negro History*, LX:2 (Apr. 1975), 257–258; McMurry, 74–75, 80.

2 GWC, *How to Build Up Worn Out Soils*,

Tuskegee Institute Experiment Station, Bulletin 6 (Tuskegee, Alabama: 1905), 4 (CM/46).

3 GWC, "Prof. Carver's Advice to Farmers: Cheap Cotton," *Colored Alabamian*, Feb. 13, 1909, clipping (CM/46); GWC to A.C. True, Sept. 18, 1902 (CM/2); GWC, *When, What, and How to Can and Preserve Fruits and Vegetables in the Home*, Tuskegee Institute Experiment Station Bulletin 26 (Tuskegee, Alabama: 1915), 3 (CM/46); Booker T. Washington, "A Farmers' College on Wheels," *World's Work*, Dec. 1906, 8354.

4 Campbell, *Movable School*, 82; GWC to Booker T. Washington, Nov. 16, 1904 (CM/2).

5 Jones, 263–264; Campbell, *Movable School*, 91–93.

6 GWC, "What Shall We Do for Fertilizers Next Year?" Tuskegee *Messenger*, Dec. 2, 1916, clipping (CM/46).

7 GWC, "Top Soil and Civilization," typescript of editorial from Montgomery *Advertiser*, June 21, 1938 (CM/61).

8 GWC, "Being Kind to the Soil," *Negro Farmer*, Jan. 31, 1914, clipping (CM/46).

9 GWC, stenographic report of lecture at the Voorhees Farmers' Conference, Voorhees Normal and Industrial School, Denmark, South

Carolina, Feb. 16, 1921 (CM/46). Carver's statement that during the Jubilee year, "Nothing must
be taken off of [the soil]" does not strictly accord
with the biblical passage that apparently served
as his inspiration, Leviticus 25:11–12. In the New
Revised Standard Version, the passage reads as
follows: "That fiftieth year shall be a jubilee for
you: you shall not sow, or reap the aftergrowth, or
harvest the unpruned vines. For it is a jubilee; it
shall be holy to you: you shall eat only what the
field itself produces."

10 James H. Cobb Jr., "Ford and Carver Point
South's Way," Atlanta *Journal*, Mar. 17, 1940, clipping (CM/62).

CHAPTER 4: "A GREAT TEACHER"

1 Thrasher, 163–165; Harlan, *Booker T.
Washington: The Making of a Black Leader,
1856–1901*, 198–199.

2 Jones, 255–256.

3 Ibid., 259–260; GWC, "Twelve Reasons Why
Every Person in Macon County Should Attend the
Macon County Fair," n.d. (CM/46).

4 Jones, 261; McMurry, 118–119.

5 McMurry, 105–106, 119; John Griffith, quoted
in Clement Richardson, "A Man of Many Talents:
George W. Carver of Tuskegee," *Southern
Workman*, Nov. 1916, 603; Booker T. Washington to

GWC, Feb. 26, 1911 (CM/4); Anonymous GWC biographical sketch, enclosure to Jesse O. Thomas letter to GWC, Apr. 11, 1918 (CM/5); .

6 GWC, "The Need of Scientific Agriculture in the South," *American Monthly Review of Reviews*, Mar. 1902, 321; "Tenth Annual Bible Conference," Albany Institute *News*, Mar. 1915, clipping (CM/60).

7 "Macon County Fair," Tuskegee *Student*, Nov. 13, 1915, clipping (CM/60).

8 GWC, *Feeding Acorns*, Bulletin 1, Tuskegee Institute Experiment Station (Tuskegee, Alabama: 1898), 5 (CM/46); William R. Carroll and Merle E. Muhrer, "The Scientific Contributions of George Washington Carver" (unpublished report, prepared for the National Park Service, 1962), 2, 10–11, 19–22; Barry Mackintosh, "George Washington Carver: The Making of a Myth," *Journal of Southern History*, XLII:4 (Nov. 1976), 508–510, 514.

9 GWC, *Feeding Acorns*, 4; GWC to A. Carmack, May 2, 1942 (CM/41).

10 GWC, speech presented Sept. 14, 1920, to a convention of the United Peanut Association of America held in Montgomery, Alabama, printed in an unidentified publication, clipping (CM/60); G. Lake Imes, *I Knew Carver* (Harrisburg, Pennsylvania: J. Horace McFarland Company, 1943), 8 (CM/59).

11 GWC, "Nature as Our Greatest Educator," 217; Roscoe Dunjee to GWC, Feb. 2, 1936 (CM/18).

12 GWC, "A Few Hints to Southern Farmers," *Southern Workman*, Sept. 1899, 352.

13 GWC, *Nature Study and Gardening for Rural Schools*, Bulletin 18, Tuskegee Institute Experiment Station (Tuskegee, Alabama: 1910), 3 (CM/46).

14 GWC, "What Chemurgy Means to My People," 41.

15 GWC, "Prof Carver's Advice to Farmers: Cheap Cotton"; Richardson, 600.

CHAPTER 5: "WALKING ON ROSES"

1 Guzman, notes of interview with Etta M. Budd; GWC, "Nature as Our Greatest Educator," 217.

2 GWC, *Some Ornamental Plants of Macon County, Ala.*, Bulletin 16, Tuskegee Institute Experiment Station (Tuskegee, Alabama: 1909), 5, and *Twelve Ways to Meet the New Economic Conditions Here in the South*, Bulletin 33, Tuskegee Institute Experiment Station (Tuskegee, Alabama: 1917), 7 (both on CM/46).

3 Glenn Clark, *The Man Who Talks With the Flowers* (St. Paul: Macalester Park Publishing Company, n.d.), 40, and Imes, 9–10 (both on CM/59).

4 GWC, *White and Color Washing with Native Clays from Macon County, Alabama,* Bulletin 21, Tuskegee Institute Experiment Station (Tuskegee, Alabama: 1911), 5–6 (CM/46).

5 "Black Leonardo," *Time,* Nov. 24, 1941, 82.

6 Clark, 20.

CHAPTER 6: "TWO OF THE GREATEST PRODUCTS THAT GOD HAS EVER GIVEN US"

1 Washington to GWC, Feb. 26, 1911; McMurry, 53–70, 162.

2 GWC, stenographic report of lecture at the Voorhees Farmers' Conference.

3 Clarence T. Mason, "George Washington Carver," *Chemurgic Digest,* Aug. 1966, 4.

4 "Hearings Before the Committee on Ways and Means, House of Representatives, on Schedule G, Agricultural Products and Provisions, January 21, 1921," *Tariff Information, 1921* (hereafter cited as *Tariff Information*) (Washington: Government Printing Office, 1921), 1544 (CM/60); "Wonderful Possibilities in Southern Crops," *Modern Farming,* Oct. 10, 1920, clipping (CM/60); GWC to Robert R. Moton, Jan. 29, 1918 (CM/67); anonymous GWC biographical sketch, enclosure to Jesse O. Thomas letter to GWC.

5 GWC to Robert R. Moton, Sept. 22, 1919

(CM/6); *Tariff Information*, 1545, 1547–1548.

6   GWC, speech presented to convention of United Peanut Association of America; *Tariff Information*, 1543–1551.

7   Wheeler McMillen [writing under the pseudonym W.W. Wheeler], "'Great Creator, I Said, Why Did You Make the Peanut?'," *Farm and Fireside*, Nov. 1928, 8.

CHAPTER 7: "GOD SPEAKING TO MAN THROUGH THE THINGS HE HAS CREATED"

1   Clark, 17, 21.

2   Frank H. Leavell, "George Washington Carver: An Interview for Students," *Baptist Student*, Nov. 1938, 6 (CM/59). Carver's biblical account is drawn from Genesis 1:1 and 1:29, but as quoted by Leavell, it varies slightly from the wording of the King James Version, which is Carver's apparent source. The actual wording in the King James Version is as follows: "In the beginning God created the heaven and the earth." (Genesis 1:1) "And God said, Behold, I have given you every herb bearing seed, which *is* upon the face of all the earth, and every tree, in the which *is* the fruit of a tree yielding seed; to you it shall be for meat." (Genesis 1:29)

3   GWC, "The Love of Nature," *Guide to Nature*, Dec. 1912, 228.

4 "Negro Scientist to Revolutionize Three Food Crops," Associated Press story (name of newspaper obscured on microfilm image), Nov. 19, 1924, clipping, and "Men of Science Never Talk That Way," New York *Times*, Nov. 20, 1924, clipping (both on CM/60).

5 GWC, Iowa State College *Alumnus*, Apr. 1925, 239–240, clipping (CM/60).

6 Charles Henry East, "An Unscientific [?] Scientist," *Golden Age*, July 15, 1925, clipping (CM/60). (Brackets are in original title.)

7 Christy Borth, "My Last Visit With Dr. Carver," *Chemurgic Digest*, Apr. 29, 1944, 111. Borth incorrectly gives 1940 as the year of Carver's visit to Starr Commonwealth.

8 B.B. Walcott, "Meet George Washington Carver, American Artist," *Service*, Jan. 1942, 30.

CHAPTER 8: "THE FIRST AND GREATEST CHEMURGIST"

1 For background on the chemurgy movement, see Christy Borth, *Pioneers of Plenty* (New York: The Bobbs–Merrill Company, 1939), and Wheeler McMillen, *New Riches From the Soil* (New York: D. Van Nostrand Company, 1946).

2 Wheeler McMillen, "'Great Creator, I Said, Why Did You Make the Peanut?'," 8; Borth, *Pioneers of Plenty*, 226.

3 Wheeler McMillen, "A Rubber Crop for Your Farm," *Farm and Fireside*, June 1928, 47; GWC to Henry Ford, July 28, 1941, Ford Archives, accession 285, Office of Henry Ford, box 2453.

4 GWC, stenographic report of a speech delivered at Tuskegee Institute chapel, Oct. 20, 1940 (CM/46).

5 John Ferrell, "Henry Ford's Vision of a Sustainable Future," *EarthSave*, Spring/Summer, 1992, 16–17.

6 GWC, "The Need of Scientific Agriculture in the South," 321.

7 Holt, 238.

8 "Finds Rubber Forest in Tulsa," Oklahoma City *Black Dispatch*, Oct. 13, 1927, clipping (CM/60).

9 GWC to A.A. Hyde, Mar. 5, 1930 (CM/12).

CHAPTER 9: "NATURE'S REMEDIES, WHICH GOD INTENDED WE SHOULD USE"

1 Richardson, 602; GWC to Booker T. Washington, July 27, 1914 (CM/5).

2 GWC, *Three Delicious Meals Every Day for the Farmer*, Bulletin 32, Tuskegee Institute Experiment Station (Tuskegee, Alabama: 1916), 4–6 (CM/46).

3 GWC, "A Few Notes on a Demonstration by Dr. Carver Before the Class in Commercial

Dietetics," Tuskegee *Messenger*, July–Sept. 1936, 14, clipping (CM/46).

4 GWC to Charles Freer Andrews, Feb. 24, 1929 (CM/11); GWC to R.B. Eleazer, Nov. 29, 1930 (CM/12).

5 *Tariff Information*, 1549; W.E. Tabb to GWC, Feb. 16, 1929, and GWC to Tabb, Apr. 16, 1929 (both on CM/11); *How to Grow the Peanut and 105 Ways of Preparing It for Human Consumption*, Bulletin 31, Tuskegee Institute Experiment Station (Tuskegee, Alabama: 1916), 19–20 (CM/46).

6 Ferrell, 16; "Amazing Food Uses for the Lowly Peanut," St. Louis *Globe–Democrat*, Apr. 3, 1921, clipping (CM/60); bibliographic information provided to author by William Shurtleff, Director, Soyfoods Center, Lafayette, California, Mar. 1993.

7 GWC to Henry Ford, Sept. 3, 1942, Ford Archives, accession 285, Office of Henry Ford, Box 2453.

8 Frances Moore Lappé, *Diet for a Small Planet*, tenth anniversary edition (New York: Ballantine Books, 1982), 66–88; John Robbins, *Diet for a New America* (Stillpoint Publishing, 1987), 313–381.

CHAPTER 10: "WE HAVE EVERYTHING WE NEED"

1 Austin W. Curtis Jr., "Memoirs of His Life and Work with Dr. George Washington Carver and Henry Ford," interview conducted by Dave Glick and Doug Bakken, July 23, 1979, Ford Archives, accession 1600.

2 GWC, "Alcohol as Fuel," letter published in Birmingham *News*, Aug. 8, 1935, clipping (CM/61).

3 W.T. Maynor, "Slave–Born Chemist Deplores Great Waste in American Industry," Birmingham *News*, July 27, 1941, clipping (CM/62).

4 Charles W. Greenleaf to GWC, Sept. 10, 1942, and GWC to Greenleaf, Sept. 14, 1942 (both on CM/42); McMillen, *New Riches From the Soil*, 285–295.

5 McMillen, *New Riches From the Soil*, 197–200; "Hemp, Perennial War Baby," *Industrial and Engineering Chemistry*, July 1945, 14.

6 "Chemurgy: A New and More Bountiful Era Emerges from Our Farms and Laboratories," *Newsweek*, Dec. 3, 1951, 82–83; "New Crops, New Uses, New Markets: Industrial and Commercial Products from U.S. Agriculture," *1992 Yearbook of Agriculture* (Washington, D.C.: U.S. Government Printing Office, 1993), 6, 148–150, 215–218; Lee R. Lynd et al., "Fuel Ethanol from Cellulosic Biomass," *Science*, Mar. 15, 1991, 1319.

7 Wheeler McMillen, "A Historic Perspective," in *Commercializing Industrial Uses for Agricultural Commodities*, proceedings of a conference by the same name held in Washington, D.C., Mar. 14–16, 1990.

CHAPTER 11: "I AM A BLAZER OF TRAILS"

1 Holt, 269; Leavell, 6.

2 GWC, "Grafting the Cacti," *Transactions*, Iowa Horticultural Society (1893), 257, quoted in McMurry, 38.

3 Jessie L. Guzman, script for (or transcript of) interview on unidentified radio program, Jan. 12, 1953 (CM/63); GWC, *Feeding Acorns*, 6; GWC to Grady Porter, Feb. 3, 1942 (CM/40).

4 Norman Myers, *The Primary Source: Tropical Forests and Our Future* (New York: W.W. Norton and Company, 1985), 189–259.

5 Daniel Goleman, "Shamans and Their Lore May Vanish With Forests," New York *Times*, June 11, 1991, C1, C6.

6 Lester R. Brown, *Food or Fuel: New Competition for the World's Cropland*, Worldwatch Paper 35 (Washington: Worldwatch Institute, 1980), 6; "Food versus Fuel? The Moral Issue in Using Corn for Ethanol," Wallace E. Tyner, *Technology Review*, July 1981, 76–77; Lynd et al., 1318–1323; Lappé, 69–71.

7 GWC, "The Need of Scientific Agriculture in the South," 320–321.

Also available from
Macalester Park Publishing Company
George Washington Carver
The Man Who Talks With The Flowers
$4.95 paperback

Fruits of Creation
A Look at Global Sustainability as Seen Through
the Eyes of George Washington Carver
$8.95 paperback

These and other fine books available from
Macalester Park Publishing Company
8434 Horizon Drive
Shakopee, MN 55379
1-800-407-9078

To learn more about the mission and programs
of the Green Cross, please write or call
The Christian Society of the Green Cross
10 East Lancaster Ave.
Wynnewood, PA 19096
1-800-650-6600

Additional copies of
Fruits of Creation
are available from either
The Green Cross or
Macalester Park Publishing Company